THE COSMIC CREATION

TRANSFORMATIONS

GAROUCH T. POGOSYAN

THE CREATION

There are two sources of information of the creation of the world - Religion and Science.

In Religion, the Creator is God, the Supreme Reality, The Being, perfect in power and wisdom.

In science, the Creation proceeds by physical processes.

The ancient Hindu Religious Scriptures, called Vedas, recognize the Supreme Being - perfect, absolute, infinite - which creates everything for it's delight only, without regard for the hardship of all creatures, human beings included. These Vedas have nothing to say about the purpose in or behind creation.

Some individuals developed special types of Yoga meditation to get connected with Supreme Self by reaching the highest possible consciousness - Brahmin. As a result, the following passages appeared in one of the Dharmas.

"The manifold world is created from the UNITARY soul as a creative or Self-expressive activity which has the center of its being Self-delight issuing forth, in endless forms, the fundamental fact of its existence."

Therefore, human beings, as just another creature, cannot interfere with the game of the Supreme-Self.

However, the Vedas teach that, although human beings suffer the miseries of life and death, it is transient and should just be borne, because at death, the human self does not cease to exist. It acquires a new body. Then it comes back, and a new cycle starts. To escape it, we should practice the Yoga meditations to Brahman, the supreme consummation of knowledge, abandoning the life of the senses, attachments and antagonisms. Abandon conceit, violence, lust, anger, and superfluous possessions. Be selfless and tranquil and become fit for Brahman consciousness. Then, by the grace of the Lord, attain the eternal indestructible state. The requirements are very severe, and many different Yoga meditations are very difficult. And, to attain Nirvana depends on the grace of the Creator.

Later, some 2600 years ago, a Yogi by the name of Gautama Buddha achieved the top limit of Yoga meditation, but could not experience the Supreme Being. Then, he decided to go into deeper and increasingly longer meditation to get in touch with the Supreme Self. But, he did not succeed. The only thing was a whisper: "live sir, life is good"; because he was not eating in order to kill his body and mind. He also saw that generations of human beings continually are born, suffer life, and die. They come to life and go through the cycle indefinitely. He was the son of the regional king and had all of the luxuries of life. But, he became a wanderer living only on alms, teaching people that the so-called Creator does not exist, and that human beings are the highest living beings, suffering life.

To escape the cycle of birth and death, it is enough to practice a single Yoga meditation, achieve peace and kindness to other human beings and animals. Do everything the way you think is the right way. Abandon the life of the senses, lust, anger, and material possessions.

Why was Buddha so disappointed in life?

He hoped that by communicating with the Supreme-Self, he could find the purpose of life. He did not find anything. Then, he decided not to come back to life. Did he succeed? Nobody knows.

There are also other religious theories of Creation. But, the main theory is given by Moses in the book of Genesis in the Bible.

Like in the Vedas, here, too, God creates for his delight; declaring everything good. By His will, the God Jehovah created everything out of nothing. Here is a brief description.

The first things He created were space and the Earth. Then, He created light. After that, He created every kind of plant, then the stars, the sun and the moon to shine on the earth and to serve as signs for seasons and for days and years. Next, God created fish and birds. Then, after creating all kinds of land animals, God decided to make man in His image and according to His likeness.

And Jehovah proceeded to form man of the dust from the ground and to blow into his nostrils the breath of life, and the man came to be a living soul. Then, Jehovah showed His creatures to the man. But, the man was not as happy as Jehovah was. Hence, Jehovah put the man into a deep sleep, broke one of his ribs and built a woman. They lived in the Garden of Eden, enjoying everything in the Garden. But, they knew that they couldn't eat the fruit from a certain tree, called the tree of knowledge of good and evil, as Jehovah told the man in order to preserve his life.

Since they disobeyed Jehovah's word, and ate off of the forbidden tree, the man and the woman were expelled from the Garden, and they were made to cultivate the ground for food. Later, the woman gave birth to two sons, Cane and Abel. Cane killed his brother Abel because Jehovah liked the fatty pieces of Abel's offering and ignored Cane's offering. Then, they had two other sons and daughters, who made other sons and daughters, and so on. People started to grow in numbers and the women were so good looking that the sons of Jehovah started to have sexual relations with them and producing mighty sons. And Jehovah regretted that He had created man on earth and felt pain in His heart. The earth had become filled with violence, because the evil of man was abundant on the earth.

Jehovah regretted making men and decided to wipe all living beings off of the earth. But, Noah who had three sons, found favor in the eyes of Jehovah. And God said to Noah that he would be bringing them to ruin together with the earth. He told Noah to build an ark out of the wood of a resinous tree, make compartments in the ark and cover it inside and outside with tar. The length of the ark must be 500 feet, the height 50 feet and the width 75 feet in order to withstand the flood of the rain. He told Noah to gather every kind of animal, male and female, and food for them and his family. Noah was 600 years old when it started raining for 40 days and 40 nights. All the mountains were covered with water for 150 days.

Thus Jehovah wiped out every existing thing from the surface of the earth. The ark came to rest on the mountains of Ararat, and the waters had drained from the earth. Then, Jehovah let the family of Noah and all of the animals out of the ark so that they could wander the earth and be fruitful and become numerous. Noah built altar to Jehovah and burnt offerings upon it.
Jehovah smelled the calming scent and decided to never again call down evil upon the ground on man's account, because the inclination of the heart of man is bad from youth on up. Then, Jehovah told Noah that human beings could eat every moving animal

and green vegetation, but not eat flesh with is soul, its blood. Jehovah made a covenant with Noah with the sign of a rainbow. Noah died at the age of 950 years. The sons of Noah were fruitful, so the population spread about the land.

Now, the theory of Moses of the creation everything existing calls for some brief analysis.

In the beginning, God created the heavens and the earth. There was no space until God started its activity. There was only a point, which God exploded into a vast expanse of the universe. But, if there was no space, then where was God? Was God inside that point and did he explode Himself?

To create the earth with its huge mass of matter requires primordial matter and energy. Then, there is the fact that the earth is not the oldest body in the universe. Most probably, the earth was the integral part of some huge hot body, a star or a sun. It rotates about its axis and revolves around the sun.

This much is enough to dismiss the first axiom of Moses.

And God said, let there be light and there was light.

But, in nature, light is emitted from the hot bodies of stars or suns. It cannot be created from nothing.

Then, God brought the waters together into one place so that the earth would be able to grow grass, vegetation and all kinds of fruit trees.

This would be impossible without the sun's energy. But, according to Moses, the sun was installed in heaven after the plants were growing.

After the sun, God was ready to create every kind of fish, flying creatures, land animals, and human beings.

This is impossible if God had neglected to create air. This is the sequence of the creation of all living beings. First of all, God created all vegetation and fruit trees bearing seeds for the further development of food for everything moving upon the earth in which there is life in the form of a soul.

Then, God created the fish and the flying creatures. After that, God created all kinds of animals and human beings. Every living being was made according to its kind, male and female. Further, God blessed them and said to them, "Be fruitful and become many and fill the earth and subdue it and have in subjection the fish of the sea and the flying creatures of the heavens and every living creature that is moving upon the earth.".

When Jehovah settled Adam in the Garden of Eden to cultivate it and care for it, He commanded Adam to not eat from the tree of the knowledge of good and evil. For, on the day that he ate from it, he would positively die. But, Jehovah knew that he was not telling the truth. And when Eve and Adam ate from that tree and obtained the knowledge of good and evil, Jehovah put them out of the Garden of Eden in order that they may not eat from the tree of life and live to time indefinite.

Jehovah knew that every creature, including man, is mortal, regardless of what they might eat. So, neither sin, nor death, was transferred to later generations. But, the most perplexing thing is that God created a single pair of human beings. They had many sons and daughters, who in turn had sons and daughters of their own, and so on. Who were the wives of those men? Obviously, their sisters. Jehovah enjoyed everything he played. But, how could he enjoy the sexual intercourse between those brothers and sisters?

And then, all of us are brothers and sisters.

How old were Adam and Eve at birth? What nationality were they? Were they Jewish? Moses describes Jehovah as the God of the Jewish people; He delivered Jews form Egypt, and helped to conquer all the countries in Canaan and commanded them to kill everybody, old and young, children and women, not giving them a chance to renew and redeem themselves.

Why did Jehovah create man first then hypnotize him to withdraw one of his ribs with which to make a woman? This ignores the fact that every man is born from a woman.

And how did Jehovah make only the external figure of a man, making it function without making all of the internal organs and systems?

If Jehovah was so wise and knowledgable, He should have known that every living thing is made from the combination of sperm and egg.

Since Noah was descended of a single pair, Adam and Eve, then all of his family was sisters and brothers. And, since Abraham was the descendant of Noah, then his wife was also his sister. So, all human beings are sisters and brothers.

All nations are one family.

This would be true if we were to believe that this Science Fiction story has anything to do with Reality.

The second part of this Science Fiction story is even worse.

It presents Jehovah as someone with the extreme powers of magic.

But, some theologians consider these stories to be Holy Scriptures.

Before proceeding to describe the second part of the story, I need to say something about Jehovah, our ancestors and Noah, in order to complete the first part of the story.

Moses presents his Jehovah God as a person; like a man, having all man's image and likeness, because He created man in His image and according to His likeness.

This Jehovah even had sons who made women pregnant and produced sons. Then why did He decide to make man, and then, from a man, a woman? He could have made a woman, and then produced a man by making her pregnant. That would have been more natural, because nothing comes out of man, but every male and female comes out of a female.

Jehovah was able to make the sculpture of a man and make him alive by breathing into his nostrils. But, He made only the exterior shape of a man. He had to work from the inside out to make all the interior organs and all of the complicated systems and the brains.

This single pair produced many sons and daughters. Then, the sons and daughters, having intercourse, produces their own sons and daughters. Then, their sons and daughters produced their own sons and daughters, and so on.

Was Jehovah delighted with the sexual intercourse of His sons and daughters? Did He create all nations from a single set of parents? Later, Jehovah realized that He created violent men. Given that these men had all the attributes of Jehovah, then Jehovah was also violent. But, He spared Himself and drowned all men in a great flood except for Noah, his three sons and their wives, along with pairs of every kind of living beings. But, he did not spare any plants.

Noah had to construct a tremendous sea vessel, gather pairs of every living being and food for them all for several months. Needless to say, that was impossible to do. The living beings were all over the world. Where was this vessel made? What made it flow to the mountains of Ararat?

How could the plants survive 150 days under deep waters? The global flood to cover all high mountains by rain only is absolutely impossible. The water level in the oceans had to be raised to the level of the highest mountains. Where did Jehovah get this much water?

Who created this Science Fiction, and why?

No one in his sound mind should be able to accept it.

But, for some theologians, these are the Holy Scriptures. Maybe they know secrets that are not revealed to the rest of us.

Later, when A' Bram was born in Canaan, Jehovah appeared to him and blessed him. A' Bram became the father of Isaac and, later, from his second wife, many other sons. Isaac became father to Jacob, but Jehovah changed his name to Israel. Israel became father to 12 sons.

Because of a severe famine, Israel's entire family moved to the land of Egypt. They were about 70 people. They dwelled in the very best land of Goshen. Israel died, but his sons became fruitful and began to multiply; and they kept on multiplying and growing mightier at a very extraordinary rate, so that the land got to be filled with them.

Then the king of Egypt commanded of his people to throw every newborn son into the river Nile. But, the newborn Hebrew boy, Moses was adopted by the Pharaoh's daughter and saved. As Moses grew up, he went to see how the Hebrew people were living. He saw an Egyptian striking a Hebrew man, so Moses struck the Egyptian down and hid him in the sand. When Pharaoh learned about this, he attempted to kill Moses. But, Moses ran away to the land of Median. There, he took one of the priest's daughters as his wife. Later, she bore him a son. Moses became a shepherd of the flock of the priest.

In the wilderness, Jehovah talked to Moses through a
burning bush, telling him to go and lead the Hebrew
people out of Egypt to the land of Canaan. And, by
Jehovah's instructions, Moses went to Pharaoh and
said to him that Jehovah, the God of the Hebrews,
commanded them to go to the wilderness and
sacrifice to Jehovah. It would take three days.

Thus far, there is nothing extraordinary, but now
begins the fairy tales. By the suggestion of his God,
Jehovah, Moses threw his rod on the ground, and the
rods became big snakes. Moses' snake swallowed
the other snakes.
This was not enough to sway Pharaoh's decision.
Next, Moses lifted up his rod and struck the water of
the Nile River and the water turned into blood.

Then, Moses, using his magic rod, did many other
tricks, but nothing helped, because the king of Egypt
knew that Moses had something evil on his mind.
Finally, Jehovah told Moses that He would bring
another plague upon Pharaoh and Egypt; that every
firstborn in Egypt must die, from the firstborn of
Pharaoh to the firstborn of every beast. The Hebrews
could pass over the plague and protect their
firstborns by sprinkling their doors with the blood of a
ram.

The plague was carried out, after which Pharaoh
gave Moses permission to take his people and go
serve their God. But, Moses took them out of the

country and headed for Canaan. With them, they took from Egypt articles of silver and gold and many flocks and herds of animals. When the king of Egypt learned of this, he took his army and pursued Moses and his people. When Moses reached the Red Sea, he saw that the army was approaching behind them. Stretching his hands to the sky, Moses parted the waters and led his people through the Red Sea. When the army of Pharaoh followed Moses and his people into the Red Sea, Moses lifted his staff and the waters came rushing back and covered the whole army.

By performing many other miracles, Moses was able to provide his people with cold and fresh water as they wandered the desert for forty years. Finally, they reached the border of Canaan, where Moses died. With the help of Jehovah, Joshua led the Hebrews into the city and captured it.

But, the death of Moses is written about in the book of Deuteronomy. It says that Jehovah buried him, and nobody knows the place of his burial until this day.

Moses was the highest prophet of the Hebrew people. He gave them all the laws and the rules on how to relate to God and to each other.

Although there are many difficulties presented in the handing down of the visions of Moses which were given to him in the Hebrew language, translated into Aramaic after his death, and thousands of years later, translated again into modern languages. The teachings of Moses that God was with them gave them confidence and unified them against their enemies.

But, every time they were losing their trust in God, which weakened their unity, the enemies were taking advantage and fighting against them. The first, and the most aggressive enemy was Babylon (present day Iraq). Israel was defeated and all able-bodied men and women were taken as slaves to work in Babylon. After some 400 years in confrontation with Babylon, Persia (now Iran) allowed the Israelites to return to Israel. But then, the Turks took over. Finally, the Roman Empire captured and colonized Israel.

The word Jehovah as the Cause of the Creation of the Universe is not mentioned in the Religion of Christianity.

Although in the first Gospels of Jesus there is nothing about the Creation of the universe and life, the last gospel written by John, youngest and most beloved by Jesus, declares that the Spiritual Being, who appeared to us as the human being, was the creator of everything.

"All things were made through Him, and without Him, nothing was made that was made.".

This spiritual being "Became flesh and dwelt among us, and we beheld His glory, the glory as of the only begotten of the Father, full of grace and truth.".

In the human body, this Spiritual being was called Immanuel, or, from the Greek, Emmanuel or Iesous, which means "God with us".

Another theory designates the essence of reality as Nature without her unconscious physical energy and blind mechanical causation and seeks to explain mind, spirit, value, etc., as accidental by-products of the interactions of physical forces.

The commonly accepted scientific theory of the creation of the universe is the Big Bang. It suggests that all matter of our universe was concentrated in an extremely dense and hot small ball. The ball grew through expansions and it is still expanding. But now, it expands more rapidly. Some unknown force or energy causes this acceleration of the expansion.

This conclusion is based on the observation that all galaxies have their spectrum lines of light increase in wavelengths.

This phenomenon is called the red shift of wavelengths. The scientists assumed that this red shift is the result of the Doppler effect.

The Doppler effect was observed in sound waves. When the source of sound and the receiver are moving away from one another the frequencies of sound waves decrease, and therefore the wavelengths increase.

If this is true for light also, then the galaxies are moving away from each other with velocities comparable with the speed of light, because the fractional change of the wavelength shows a good fraction of the speed of galaxies and light.

$$(L' - L)/L = V/C$$

Where L' is the wavelength of light coming to us from the galaxy and L is the wavelength of light measured in our laboratories on the earth. And V is the speed of the moving galaxy and C is the speed of light in a vacuum.

But all experiments show that light is a particle. It does not resemble the sound waves. Yet, like all subatomic fast moving particles, it makes waves in space, because space is full of extremely tiny particle; billions times billions smaller than the electron.

The frequency of these waves is proportional to the energy of the photon, the particle of light.

$$f = E/h$$

where h is Plank's constant, and E is the energy of the photon.

The photon, as any particle, has a mass, m, and its energy is $E = mc^2$. The photon is not a mass-less particle as all scientists think.

The photon coming to us from a massive galaxy loses some of its energy by the gravitational pull of the galaxy; therefore its wave frequency decreases and the wavelength increases.

It is obvious that the scientific theory of the Big Bang is a false theory.

Yet scientists state that the theory is correct because there is the background microwave radiation from every direction detected by the electronic devices which is evidence of the Big Bang and expansion of the universe. This is another false reasoning as we will find later, in the second part of this book.

Another so-called scientific theory of the creation of the universe is the Steady-State, or the continuous creation Cosmology which envisions an unchanging Universe that had beginning and will have no end, because with expansion of universe new stars are created to fill the space.
This is another false theory. I showed that the universe does not expend.

Luckily, these theories have nothing to offer about the creation of the living beings.

However, there is continuous research by the naturalists and biologists to solve that problem. These are the theories given by Darwin and many biologists. However those theories are not better than the two theories of the creation of the Universe. They describe some observations, but when they attempt to explain them and get closer to describing the origin of life, they fail.
The next pages give the descriptions of those theories.

THE ORIGIN OF SPECIES

Darwin

This confusing, contradictory and lengthy "Abstract", as he defined his writing, summarizes Darwin's work in skeletal form. However, his writing can be reduced to a few pages without reducing the context.

First of all, he does not touch the origin - the first species. Although he does not give us a definition of the word species, we know that a species is the same kind of plant or animal. What Darwin discusses is evolutionary theory, because he describes how species are modified from generation to generation. He also states that his theory does not apply to human beings, but we will see that in a single sentence, he contradicts himself.

Darwin's main principles state that each species has not been independently created, but that every species is a lineal descendant of some other species. He also stated that natural selection has been the most important, but not the exclusive, means of modification. He uses the term natural selection in analogy of the human selection of the best, those who most easily adapted to the living conditions and were the fittest to survive. But, then, he writes:

In the literal sense of the word, no doubt, natural selection is a false term. He says that it stands for the term survival, and is used for brevity.

Another of Darwin's principles is the struggle for existence, which is only common sense.

Then he explains that his theory is a theory of evolution. He writes:

Almost every part of every organic being is so beautifully related to its complex condition of life that it seems as improbable that any part should have been suddenly produced perfect, as that a complex machine should have been invented by man in a perfect state. Every species and their parts are produced in less and simple imperfect state, and tend to modify to achieve the perfect state as we observe now. These modifications are slow and small quantities, but during the many generations, they add up by the principle of inheritance and the results in the improvement of part and structure of the species.

The similar framework of bones in the hand of a man, wing of a bat, fin of the porpoise, and the leg of the horse, and innumerable other such facts, at once explain themselves on the theory of descent with slow and slight successive modifications. On the principle of successive variations, not always supervening at an early age, and being inherited at a corresponding not early period of life, we clearly see why the embryos of mammals, birds, reptiles and fishes should be so closely similar, and so unlike the adult forms. We may cease marveling at the embryo of an air-breathing mammal or bird having branchial slits and arteries running in loops, like those of a fish, which has to breathe air dissolved in water by the aid of well-developed branchiae.

There is direct evidence that some progenitor species has given birth to other and distinct species. I believe that animals are descendants from, at most, only four or five progenitors, and plants from an equal or lesser number. Or, all animals and plants are descended from some one prototype. All living things have much in common, in their composition, their cellular structure, their laws of growth, and their liability to injurious influences. With all organic beings, excepting perhaps some of the very lowest, sexual reproduction seems to be essentially similar. With all, as far as is at present known, the germinal vesicle is the same so that all organisms start from a common origin.

As Professor Asa Gray has remarked, "the spores and other reproductive bodies of many of the lower algae may claim to have first a characteristically animal, and then an unequivocally vegetable existence." Therefore, it does not seem incredible that, from such low and intermediate form, both animals and plants may have been developed. All the organic beings which have ever lived on this earth may be descended from someone primordial form. But, this inference is chiefly grounded on analogy, and is immaterial whether or not it be accepted. No doubt it is possible, as Mr. G. H. Lewes has urged, that at the first commencement of life, many different forms were evolved; if so, we may conclude that only a very few left modified descendants. For, as I have recently remarked in regard to the members of each great kingdom, such as the Vertebrata, Articulata, etc, we have distinct evidence in their embryological, homologous, and rudimentary structures, that within each kingdom all the members are descended from a single progenitor.

Authors of the highest eminence seem to be fully satisfied with the view that each species has been independently created. To my mind, it accords better with what we know of the laws impressed on matter by the Creator. The production and extinction of the past and present inhabitants of the world should have been due to secondary causes, like those determining the birth and death of the individual. I view all beings not as special creations, but as the lineal descendants which lived long ago. Judging from the past, we may safely infer that not one living species will transmit its unaltered likeness to a distinct futurity. And, of the species now living, very few will transmit progeny of any kind to a far distant futurity, for the manner in which all organic beings are grouped, shows that the greater number of species in each genus, and all the species in many genera have left no descendants, but have become utterly extinct.

As all the living forms of life are the linear descendants of those that lived long ago, we may feel certain that the ordinary succession by generation has never once been broken, and that no cataclysm has desolated the whole world. Hence, we may look with some confidence to a secure future of great length. And, as natural selection works solely by and for the good of each being, all corporal and mental endowments will tend to progress towards perfection.

It is interesting to contemplate a tangled bank, clothed with many plants of many kinds, with birds singing on the bushes, with various insects flitting about, and with worms crawling through the damp earth, and to reflect that these elaborately constructed forms so different from each other, and dependent upon each other in so complex a manner, have all been produced by laws acting around us.

These laws, take in the largest sense, being growth with reproduction; inheritance which is almost implied by reproduction, variability from the indirect and direct action of the conditions of life, and from use or disuse; a ratio of increase so high as to lead to a struggle for life, and as a consequence to natural selection and the extinction of less improved forms, from the war of nature, from famine and earth, the most exalted object which we are capable of conceiving, namely the production of the higher animals, directly follows. There is grandeur in this view of life, with its several powers, having been originally breathed by the Creator into a few forms or into one. From so simple a beginning, endless forms most beautiful and most wonderful have been and are being evolved.

Darwin did his research work in the 1840's-50's. His primary interest was to find the origin of life; plants and animals. Somewhat confused, he made a sickening religious conclusion; there was something there, having potential animal and plant characteristics. The Creator breathed His life into the nostrils of that thing and it came to life, which gave rise to all species of animals and plants.

But, Darwin forgot to include both the male and female potentials in that thing.

Many people, whether or not they are intellectuals, are interested in learning the secrets of life. After almost one hundred years of Darwin's research, there was general acceptance that genes were special types of protein molecule. In 1944, the physicist Schrodinger, in his book, *What is Life?*, expressed his belief that genes were the key components of living cells.

To understand what life is, biologists must find out how genes act. Using crystallographic methods, biologists determined the three-dimensional double helix structure of DNA, which occur in the chromosomes of all cells, in the nuclei of the cell. And all genes are composed of DNA molecules. So, the molecules of DNA determine how genes act (1959).

Later research, using more advanced devices and methods, reveal many more things in detail in the cells. Observing all of those things, biologists try to convince us that all processes in cells can be explained by physical and chemical laws. The specific structure and arrangements within and between molecules make the cells alive. And the cells are able to grow and even split into two identical cells to keep themselves alive.

Numerous experiments show that any living cell or even tiny microorganism has only a limited life span. But, kill them with strong and prolonged radiation or complete dehydration, or keep them for an extended time in a vacuum, then freeze them in liquid nitrogen 321 degrees below zero, regardless of how long, after returning, they show none of the characteristics of a living cell or organism. A tiny drop of water brings them back to life.

The actions of water can also be explained by physical laws, say the biologists. So, there cannot be any supernatural force acting here. There is the possibility that all living beings started from the single species; algae, bacteria, virus or something else, nobody knows.

How correct are the teachings of our contemporary biologists?

There is a general and single principle; one living being starts life from the combination of the male sperm and the female egg. But, there are a few methods of doing it. Fish do it in one way, birds do it another, and mammals do it in a straight way. Plants do it in a completely different way.
Second. Killing cells and microorganisms, freezing them and bringing them back to life.

Take a few days old human sperm and some very fresh sperm and repeat your experiment. You can revive the fresh sperm, but cannot bring the old sperm back, because you froze them when they were really dead. Or freeze some cells from a dead body. You can never bring them back, because they were really dead. Or, you can freeze the dead skin cells, which are on the surface of the living skin cells. They cannot be brought back to life, because they were really dead.

The conclusion is straightforward. The revived cells before freezing were the ones that were alive. Now, why does freezing keep them alive?

Every scientist accepts the concept of energy; potential and active — very different types. How can you prove that there is no potential and active life energy?

If any type of potential and active energy can be transformed into one another, then why can't the same thing be done with life energy?

But how about a drop of water? It just activates the potential life energy.

Can we detect and measure some type of potential energy without transferring it into active energy? The same is true with life energy.

Third. Following Darwin, biologists made the conclusion that all living beings started from the single type of thing, probably a virus; male, female, plant, and animal, dead and alive at any instant. So, here the virus replaces Jehovah, the God of Moses and the Creator of Darwin.

It is true that the virus is the strongest thing in the universe. But, it is the secondary creature; it exists in the living cells of bacteria or other living beings.

Normally, our immune system produces enough antibacterial bodies to kill them. Then the other system expels them from the body. The laws of physics cannot explain this complex process.

Sometimes, bacteria concentrates in a certain part of the body. The immune system sends most of its reserves to that region and leaves some other part of the body open to the overwhelming number of bacteria. The bacteria might win the war if no help was provided from the outside; antibiotics and radiation.

These parasites cannot be a prototype species. Besides, it is a biological fact that the different species do not mix to create other kinds of species. The process of coupling is strictly controlled. There are many kinds of viruses, but they are always viruses, unable to change into different species. The original species were more numerous that we observe now; some did not adapt to the environment and became extinct. Biology cannot give us the origin of life as physics cannot give us the origin of electrical charges.

THE COSMIC
TRANSFORMATIONS

Analyzing all the religious and scientific theories of the creation of the universe and living beings, we find that there exists not a single definite and convincing theory.

The religious theories have no sound basis, they are the creations of the human mind. Whoever believes it has no difficulty. The scientific theories are based on the false understanding of the real processes.

It appears that the concept of the creation of anything is false itself.

The second part of this book gives an alternative description of the Cosmic processes.

The Cosmos consists of matter and energy which can not be created or destroyed. They can only be transformed from one state to another state or from one form to another form.

At the particle level matter displays a mysterious behavior.
The particles are the electron, proton and neutron.
The electron and proton are two opposite types of matter, said to be charged in opposite ways; positive and negative.
The neutron is uncharged, and decays into the electron, proton and neutrino.
The neutrino is a neutral particle of the negligible mass.
The appearance of this tiny particle indicates that the neutron, proton, and electron consist of the flux of the tiny particles.
But the negative flux of the electron and the positive flux of the proton would explode under the repulsive force, yet they are stable. This can be possible if only these tiny particles are revolving about the common axis in one direction with the speed of light, creating a magnetic force of attraction.
This is reasonable because they have magnetic moments and act like bar magnets. This is why the electron in the vicinity of the proton flies toward the proton but stops at some distance by the repulsive magnetic force. At that small distance they start disintegrating and emit some of their combined mass into the space. The revolving speed of the revolving

particles transfers into the linear speed of the emitted combined mass of negative and positive particles. This way the photon is produced.

The energy of the emitted combine mass m is equal to $E = mc^2$. The binding energy of the electron and proton is also equal to mc^2.

The combination of the proton and neutron produces much more energy, and creates much stronger binding energy, and the emitted photon. Although light is a neutral particle, it consists of positive and negative particles, demonstrating electrical and magnetic properties.

Light is an electrical dipole.

Here I remind the readers of these properties of matter because they play a significant role in the Cosmic processes.

Now, about the energy of the Cosmos.

The energy is the power that activates matter.

The energy and matter are the properties of the Cosmos.

The combination of the electron with the proton, the neutron with the proton, the decay of the neutron is the principal method of all the transformations of matter from one state to another state.

The combination of the electron and proton produces an atom of the hydrogen, the combination of the proton and neutron produces the nucleus of the heavy hydrogen, and the decay of the neutron produces the particles electron and proton.

All these processes proceed automatically without expending any energy.

The decay of the neutron is a very important transformation, and is the starting process of all transformations of matter from one state to another state.

We observe that all matter is concentrated in the stars and other heavenly bodies. But the space between them is almost empty.

This indicates that the stars are formed from the matter of the surrounding spaces. And since their matter consists of the electron, proton and, neutron and the proton an electron are produced by the neutron, then it is obvious that the neutron was the first particle in space.

But since the neutron is made of the flux of infinitely small particles having oppositely electrical charge, then some time ago the space was occupied evenly with the oppositely charged tiny particles whirling in all directions with the speed of light.

This is the original matter and energy, which underwent all kind of transformations.

Through the electromagnetic interactions the tiny particles of space came together and formed the neutrons. The neutrons produced the electrons and protons. The electrons in combination with the protons formed the hydrogen atoms. The protons with the combination of neutrons produced the nucleus of heavy hydrogen.

Since all these processes produce energy the space was the mixture of particles and energy.
There was always enough energy to make more heavier particles, until the Uranium atoms.
Through the electromagnetic interactions all formed atoms gathered together in some region of the space and formed a huge body of the star, leaving the surrounding space with the rarefied tiny particles.

Because of the decaying processes of many nuclei continuously, energy is produced inside of star. For example, the energy of 17.2 meV is produced when lithium decays into two helium atoms upon bombardment by protons. This process produces the helium and lots of energy in the sun. Then the sun radiates some energy into the space.

According to the current theory of Cosmology the universe is about 13-17 billion years old.
The determination of the age of radioactive elements from the concentration of their decay products in minerals on our earth, as well as from outer space meteors, indicates that the time of the formation of atoms should have been about 3 to 7 billions of years ago.
The background microwave radiation which is detected from every direction of space is the combination of oppositely charged space particles which is the electromagnetic radiation, absorbed by our electronic devices.

However, we do not see the galaxies of the stars as we observe them now.
Since light takes time to reach us, we see everything in the past.
The farther we look, we see the younger galaxies of the stars.
We observe some objects so far from us that they look as bright as the galaxy of stars because we see them in their prime time.
The closer we look the older we see the galaxies, like our galaxy.

Matter, a substance seems to be a simple thing, but at close look it displays its mysterious nature. It posses two opposite powers, which are called positive and negative electrical charges.

Electric is the Greek ward amber, which shows attractive or repulsive force when rubbed with something.
Magnet is also the Greek ward, a stone, which also displays these forces.
So electric and magnetic mean amber and stone.
But we adapted these wards and think we understand what we are talking about.
Energy, the power, which activates matter is defined $E=mc^2$. But since space tiny particles always have the speed of light c, their energy is $E=mc^2$. Since the electron, proton and neutron are produced by the

combination of these tiny space particles, their rest energy is $E=mc^2$.

If the particle obtains motion energy, it gains additional mass m, and additional energy $E=mc^2$.

All these equations $E=mc^2$ are called the equations of Einstein who had nothing to do with it. But it doesn't matter.

The Cosmic Life energy is the power which makes the living things breath automatically without effort.

Animals breath in the oxygen and out the carbon dioxide. Their cells obtain oxygen and get rid of some carbon. This keeps than alive.

The cells of the plants get carbon dioxide and give away oxygen. So they mutually support one another.

All cells of a living things perform a particular functions.
But the cells which are produced by the reproductive glands of male and female bodies are the source of a new living being.

Going back in time we will see that the original living beings were also made this way. The Cosmic power provided everything, the environment conditions and the right molecules.

Using the computer language we can say that every hardware is provided with its software - the program of functioning.

The Spiritual aspect of the Cosmic Power is more illusive and difficult to accept. But it is the fact of reality.
This was revealed and demonstrated by Jesus in the first century.

Since Jesus used parables and figurative language to reveal the meaning of the kingdom of the Holy Spirit on the earth and His relation with the Holy Spirit, even His disciples were not able to comprehend its meaning.
Finally He revealed the secret to His disciples that He was using a figurative Language, and in plain language said He is the embodiment of the Holy Spirit.
He came forth from the Holy Spirit and is going back to the Holy Spirit.

Some extraordinary miracles, assigned to Jesus, turning water into wine, feeding thousands of people with a couple of loaves of bread and fish, walking on water, and so on convey the spiritual messages.
Jesus was not a miracle worker, like Jehovah with the rod of Moses.
Jesus was the Master Teacher of the Power of the Cosmic Spirit, the Holy Spirit.

Full of knowledge of the Holy Spirit, knowing that the Spirit and Life are eternal, Jesus never spoke of the Cosmic Creation and His participation in the Creation.

In his gospel John borrowed his brief theology from Moses for no apparent reason; probably to impress the religious Jewish leaders with the authority of Jesus Christ.

Nether the Holy Spirit or Jesus are the Creative sources of the Cosmos.

Jesus introduced His teachings to His disciples in His Sermon on the Mount.

It is described in the chapters of Matthew/ 5, 6, and 7.

Here the Lord's Prayer contains the core of His teachings.

'I have come into the world, that I should bear witness to the truth.

Everyone who is of the truth hears My voice.'

Jesus knew everything, past present and things to come.

Jesus knew all man, "and had no need that anyone should testify of man, or what was in man.".

"LORD, You know all things, You know that I love You." (Peter).

The all knowing Holy Spirit in Jesus gave Him this knowledge.

Jesus was the truth; He was the Holy Spirit in the human body.

The Holy Spirit is the truth.

Knowing how much the Holy Spirit loves the humans,
Jesus called the Holy Spirit our Father and His
Father.
Since the Holy Spirit was always with Him, Jesus
also called Himself the Father. "I am not alone,
because the Father is with Me."
"He who has seen Me has seen the father."
So Jesus is the only authoritative person, there is no
trinity.

After revealing the Holy Spirit as our loving Father,
and enforcing it by His healing every sickness, Jesus
explained that our Father knows all our difficulties,
and forgives us for our wrongdoings.
Since nobody has ever seen or had an experience
with the Holy Spirit it would be impossible to believe.
It would be enough to believe in Jesus, because He
is the embodiment of the HOLY Spirit.

Faith in Jesus is the first doctrine of His teachings.
Love him is the second. Here love is trust. Then
comes hope, without doubt.
So faith, love, and hope are main principals of the
teachings of Jesus.

But if the Holy Spirit, our loving and forgiving Father
provides us our spiritual needs and should make us
cheerful and happy, then why there are so many
sorrows?

The truth is, there is also the opposing spirit, which by its nature does not like humans. He makes humans act out his desires; hate one another, not trust one another, enjoy life fully while you have a life. Know that your neighbor envies you and wants your possessions. Make others serve you and never help anyone in return, if you do someone good he will return you bad, etc.

Humans unconsciously act this way and create every kind of sorrow.

All these things weaken the immune system and make them SICK.

Jesus also demonstrated the Power of the Holy Spirit by His Resurrection after the leaders of temple killed Him, because His followers were many, and they were losing their profit.

After His resurrection Jesus met with His disciples and encouraged them to continue His work.

In 1976 Jesus granted a few seconds of His presence to me face to face, gave me a vision, and fulfilled it.

This event forced me to believe in the power of the Holy Spirit in Jesus.

It is obvious that all Cosmic Powers are real; they can be observed directly or indirectly, and be experienced by some individuals.

With all its complexity, the Cosmos is a unified system; functioning according to the properties of its Powers.

It is self organizing, self content, and all transformation processes are directed towards stabilizing the system.
It is as infinite and eternal as Space is; never created or destroyed.

It is not reasonable to invent some abstract super Cosmic being, God the Creator for of everything existing.
The abstract scientific theories of the creation are only the source of confusion.

Human knowledge depends upon two factors, experience and our own intelligence. Experience is a collection of data of sense which gets their coherence solely through reasoning using our intelligence, which interprets experience.

however, wrong interpretation results in wrong understanding and therefore wrong theory.

But through analysis of the subject can find the false reasoning, and wrong theory yet the authors of the false theories of cosmic creation insist on their validity.

Moreover, most of the scientists defend and even praise the obviously false theories of.physics such as: Maxwell's wave theory of electromagnetic radiation and light, Einstein's special and general relativity, massless particles of light, Hubble's expansion of the universe, and many more theories.

Many rejections of those theories can be found in my previous book entitled *The False Theories of Physics*.

www.ingramcontent.com/pod-product-compliance
Lightning Source LLC
Chambersburg PA
CBHW051225170526
45166CB00005B/2039